# How to train your AI

by Scott Stewart

## Make It Think Like You, Not Like a Machine

First edition

ISBN:978-1-7645484-0-3

Published by Scott Stewart Publishing

Printed in Australia

**Disclaimer**

This book provides general information and practical techniques for improving the quality of written outputs generated with AI tools. It does not constitute financial, legal, regulatory, accounting, investment, or professional advice.

The examples used throughout, including references to marketing strategies, capital allocation proposals, governance documents, or board materials, are illustrative only. They are not templates for compliance, and they are not a substitute for professional judgement.

You remain responsible for the accuracy, completeness, and appropriateness of any document you produce using these techniques. Before making financial, legal, strategic, or operational decisions, you should obtain advice from appropriately qualified professionals who understand your specific circumstances.

The author accepts no liability for any loss, damage, or consequence arising from reliance on the contents of this book.

Contents

# Introduction

We have all been there. You have a tight deadline to deliver a report or strategy and your draft is messy. You feel the flow isn't right and there are some gaps in how the logic is presented. Getting the right messaging is important and the deadline is looming but the slides are only half finished. All the thinking is in your head but not fully on the page. You don't need a consultant, and you don't need theory. You just need someone to help you get your own thinking into shape without rewriting it three or four times afterwards. That's what most of us want from AI. Not brilliance. Not reinvention. You want a ghost writer that thinks and sounds like you. And when it doesn't, when it comes back with conceptual fluff, consultant rhetoric, polished symmetry, the frustration kicks in. You read it and think, what is this? Why am I getting this AI tone? Why does it sound like a generic strategy document from 2010? This has AI written all over it. I have hit that wall repeatedly.

The turning point came when I stopped blaming the output and started thinking about the mechanism and the way that it builds responses. The system is not choosing tone. It is calculating word probability. When you ask for a marketing strategy or a product roadmap, it is predicting the statistically most likely next word given the instruction and everything it has seen before. And what has it seen before? Millions of strategies. Mostly abstract. Mostly polished. Mostly structured around tidy pillars and balanced contrasts. So, when you say, "Write a marketing strategy," the gravitational pull is firstly towards what is common. It is not trying to annoy you; it is following probability gravity. If you want something else, something grounded, specific, defensible, you cannot ask for it vaguely. You must narrow the field of acceptable next words. You must remove common rhetorical pathways and force it into a corridor that looks more like your actual thinking. For example, I might ask AI to put my ideas into a logical order, and after I outline these ideas in detail, I will then give AI a page or two of my style rules that it must follow. That is necessary to build that corridor which is what this book is about. This is more important than simply pasting one giant prompt at the end of the book that worked for someone else. By understanding how the mechanism actually works and training it step by step, You can build your own library of style rules. By the end, you will not just have your own rules. You will have a working method.

# Rule 1: Concept introduction discipline

In most published writing, concepts are introduced cleanly and then explained. "A strong value proposition is essential." "Effective segmentation drives growth." "Clear positioning differentiates the brand." These patterns are everywhere. So, when you mention a concept like "value proposition" or "customer trust," the system predicts the familiar explanatory arc. That is why you get paragraphs that sound like business textbooks.
For example, if you send to AI: "*Customer trust is critical to the success of our subscription product.*" You are likely to get something back like: *"Customer trust is not a soft variable in a subscription model; it is the operating substrate. In a one-off transaction, trust influences conversion. In a subscription business, it governs survival."*

And what will follow are multiple paragraphs of academic theory. That sentence could appear in any industry, in any decade and it is statistically safe. But it probably isn't what you actually need. Perhaps it isn't the way you would normally speak, nor how others would be expecting you to speak. It may not be the style you want in your work. However if you are writing about customer trust in a real product strategy you probably know that trust shows up somewhere concrete. It shows up in refund rates, cancellation reasons, complaints about billing clarity and in review comments. So instead of allowing the system to define trust, you force it to ground trust with this: "*In the first three months after launch, 22% of cancellations cited confusion about billing cycles. The FAQ page does not explain the auto-renewal terms clearly. Refund requests increased by 14% in May.*" So now we have customer trust that is no longer a concept floating in air but it is tied to behaviour and documents; you have started to narrow the corridor.

**How to train your system**
Weak prompt:
*"write a product strategy for improving customer trust."*

Stronger prompt:
*"write a product strategy for improving customer trust. do not define 'trust' in abstract terms. anchor each claim to:*
*– a specific data point (e.g. churn rate, complaint volume),*
*– a document (e.g. website copy, pricing page),*
*– or a recorded customer insight.*
*if a paragraph reads like a textbook definition, rewrite it.*

**Add to your rule stack:**
- *No abstract concept without a concrete example, number, or document.*
- *If the sentence could apply unchanged to another company, make it specific.*

# Rule 2: the artefact trigger rule

In strategy writing, words like "shifted," "evolved," "accelerated," and "embedded" are common. They suggest progress and change. They are narratively convenient. Because they appear frequently in business commentary, the system predicts them readily. You write: *"Our go-to-market approach has evolved"*. The system continues with high-level language because that is what usually follows. But it is not what you actually need. If your go-to-market approach evolved, something changed in reality. Perhaps the pricing model changed from flat fee to tiered subscription. The target segment narrowed from "all SMEs" to "professional services firms with 10–50 employees." The ad spend shifted from television to paid search. You may have an artefact that can provide additional data points to support the analysis of what changed. So instead of: *"Our go-to-market approach has evolved."* You want: *"in April, the pricing model moved from a single $29 plan to three tiers ranging from $19 to $59. the campaign brief narrowed the primary segment to professional services firms. television advertising was paused and budget reallocated to paid search."*
Now evolution is visible.

***How to train your system***
Weak prompt:
*"Describe how our marketing approach evolved over time."*
Stronger prompt:
*"Ddescribe how our marketing approach evolved over time. if you use words such as 'shifted', 'evolved', 'accelerated', or 'embedded', you must immediately reference:*
*- a specific campaign change,*
*- a pricing update,*
*- a document revision,*
*- or a budget reallocation.*
*do not describe change without specifying what changed."*

**Add to your rule stack:**
- Movement words require evidence.
- After any claim of change, answer: what specifically changed and where is that recorded?

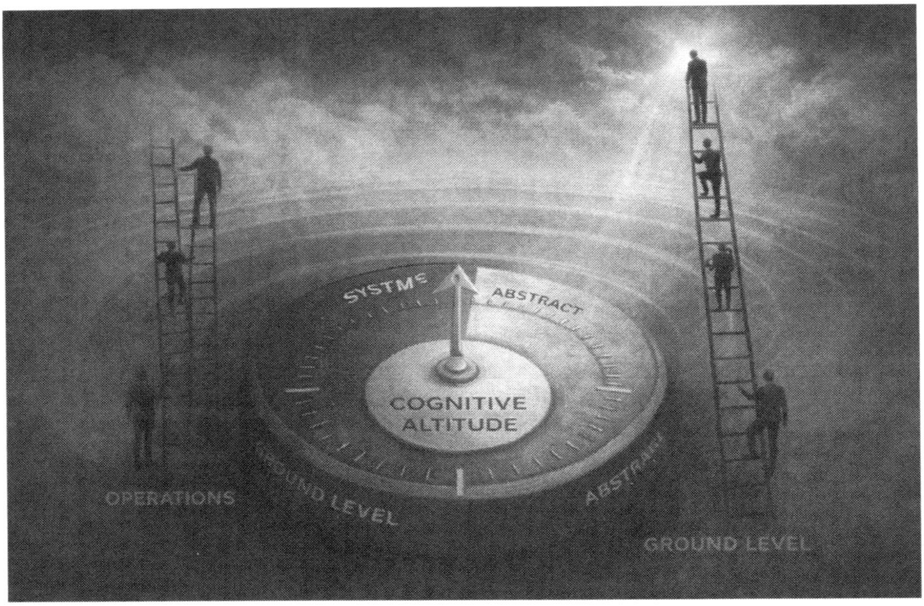


In this section I also introduce an important rule called **"Set Cognitive Altitude"**. Before you draft, decide the reading level of the output. Do not leave this to chance. The model will default upward. It will drift into abstraction, symmetrical constructions, and consultant phrasing unless constrained. If you do not specify the cognitive altitude, you are implicitly accepting its default. This is not about intelligence. It is about processing load. When I was writing staff information packs around AI the early drafts read like internal strategy documents. They were technically correct, full of qualifiers and very defensible in tone. But they required effort to absorb and understand. After I drafted all the content I sent AI a simple instruction: *"assume the audience reads at the equivalent Grade 6 level at school"*

The output changed immediately. Sentences shortened and the abstract nouns were replaced with concrete descriptions. Claims were forced into operational language. People reading the pack said, without prompting, that it was easy to understand. Nothing had been dumbed down. The thinking was the same, but the friction was much lower. Why Grade 6? In the US the current guidance for government material is to stay within the grade 6 to 8 level. Health literacy often targets grade 6 because stress reduces comprehension capacity. This can be an important consideration for the artefact you are developing. Governments have been adopting "plain English doctrine" for years now as there is a strong argument that clarity improves compliance, reduces litigation and speeds decisions. So if you want to send a clear and easy to understand message then "cognitive altitude" needs to be in your rule stack. Especially if you are drafting:

- staff information or a consultation pack,
- community communication,
- board summary,
- an executive briefing for time poor executives.

Then you should lower the altitude.
Specify something like: *"When responding, assume the audience reads at Grade 6 level. Use plain English. Prefer conversational sentences. Avoid abstract terms. If a sentence requires rereading, then rewrite it."*

If you are drafting: a commercial contract, a technical design artefact, a dense book chapter intended for a specialist reader, then you may raise the altitude deliberately but you need to consider the following as well. If lowering cognitive altitude constrains abstraction density and sentence complexity, raising it should be just as deliberate. But raising altitude is not the same as inviting flourish. It is increasing conceptual compression without inviting rhetorical inflation. That distinction matters. If you simply say: *"write at postgraduate level"* The model will often respond with ornamental vocabulary, longer sentences, and performative tone. That is not altitude. That is theatre. If you want to raise cognitive altitude properly, your prompt should control for: conceptual depth, tolerance for abstraction, sentence length, assumed domain knowledge, prohibition of simplification.
An example prompt that raises altitude without losing discipline would look something like this: *"When responding, assume the audience is senior domain experts with postgraduate-level literacy and familiarity with the field. You may use technical terminology without defining basic concepts. Maintain conceptual density and allow longer sentences where necessary to preserve nuance. Avoid oversimplification. Prioritise precision over accessibility. Do not introduce rhetorical flourish or decorative language."*
Notice what that does. It does not ask for "smarter writing". It specifies: who the reader is, what they already know, what trade-off to prefer, what to avoid. You can go further if you need a highly analytical output:
*"Assume the audience consists of specialists capable of following multi-step reasoning. You may use layered arguments and conditional logic. Retain conceptual distinctions even if they increase sentence length. Avoid examples intended for lay explanation. Write in a formal analytical tone. Eliminate narrative simplifications."*
That is appropriate for: technical design artefacts, legal analysis, risk exposure modelling, economic argumentation, academic-adjacent work
The key is to avoid confusing altitude with verbosity. The higher altitude should result in the following: tighter conceptual clustering, greater tolerance for abstract nouns, fewer explanatory bridges, more embedded assumptions, more implicit shared knowledge.
It should not result in inflated vocabulary, symmetrical rhetorical structures, consultant cadence or pseudo-academic signalling.
To illustrate, I asked AI the same question twice but each time I applied a different cognitive altitude rule and this was the result: Prompt: *"I need to speak at a forum where I explain AI and what it means for staff."*

**Lower cognitive altitude output**

(Grade 6 rule applied)
*"AI is a tool that can help us do parts of our jobs faster. It can write drafts, summarise long documents, and find patterns in data. It will not replace everyone, but it will change how some tasks are done. This means staff need to learn how to use it well and check its work. The goal is not to*

*remove people. The goal is to free up time for work that needs judgement and experience."*

**Higher cognitive altitude output**

(Expert rule applied)

*"AI refers to a class of computational systems capable of generating language, analysing data patterns, and supporting semi-automated decision processes across operational workflows. For staff, the primary impact is not wholesale replacement but task reconfiguration: routine cognitive labour becomes partially automated, shifting human effort toward oversight, exception handling, and higher-order judgement. This transition requires capability uplift, governance clarity, and explicit accountability structures to ensure outputs are validated and decision rights remain defined."*

**What does this show us?**

The grade 6 version: short sentences, one idea per sentence, concrete verbs, minimal abstraction and direct reassurance.

The expert version: longer sentences, embedded clauses, conceptual nouns such as "task reconfiguration" and "operational workflows", implicit shared knowledge, greater compression of ideas.

Neither is "better" as they are calibrated for the audience. Cognitive altitude is not a stylistic flourish. It is a control surface.

Before you draft,you decide whether this artefact requires operational grounding or system-level synthesis. If you do not specify altitude, the model will default upward into abstraction and performance language. You need to set it deliberately by stating it in the prompt and you adjust it only with intent. Cognitive altitude is a dial that you should be considering and thinking about the audience and where the dial needs to be set.

**How to train your system**

**Lower the cognitive altitude:**

Weak prompt: *"Write this in simple language for executives."*

Why its weak: Simple is subjective. Executives are not a literacy category. The model will guess. It may shorten a few sentences. It may remove a couple of adjectives. But it has no constraint on sentence length, abstraction, or processing load.

Stronger prompt: "*When responding, assume the audience reads at Grade 6 level. Use plain English. Sentences should be conversational Prefer concrete verbs over abstract nouns. Avoid jargon and technical terms unless essential. If a sentence requires rereading to understand, rewrite it. Remove qualifiers and rhetorical padding."*

Why its stronger: It defines the reading level and it constrains structure and controls abstraction. It creates a test: rereading equals failure. The model now has mechanical boundaries, not a vibe. You can tighten it further for staff or community communications: *"Assume a Grade 6 reading level. Use short sentences. One idea per sentence where possible. Replace abstract terms with operational descriptions. Avoid passive voice. Avoid multi-clause constructions. Prioritise clarity over completeness."*

*(Note: If you say "keep sentences short" that is what you get, very short sentences I find it better with "keep sentences conversational length")*

**Raising cognitive altitude:**
Weak prompt: *"Write this at a more advanced level".*
Why its weak: Advanced invites vocabulary inflation. The model may add longer words and denser phrasing, but without increasing conceptual precision. You often get theatre rather than depth.
Stronger prompt: *"Assume the audience consists of domain experts with postgraduate-level literacy and familiarity with the field. You may use technical terminology without defining foundational concepts. Preserve nuance and conditional logic even if this increases sentence length. Avoid simplification intended for general audiences. Prioritise precision and conceptual density over accessibility. Do not introduce rhetorical flourish."*
Why its stronger: It defines the audience baseline knowledge and allows abstraction deliberately. It authorises layered reasoning yet still prohibits decorative language. It increases density without inviting performance. If you need even higher analytical altitude, for example in risk modelling or economic argument: *"Assume an expert specialist audience. Retain technical distinctions. Use layered reasoning and conditional structures where necessary. Do not provide introductory explanations. Maintain formal analytical tone. Optimise for conceptual precision rather than readability for lay audiences."* The difference in output between the weak and strong prompts will be obvious. The weak prompts change tone. The strong prompts change structure.

**Add to your rule stack:**
For lowering cognitive altitude:
- Assume the audience reads at Grade 6 level unless otherwise specified.
- Use short sentences. Prefer one idea per sentence.
- Replace abstract nouns with concrete actions.
- Remove qualifiers and rhetorical padding.
- If a sentence requires rereading, rewrite it.

If you want it tighter and closer to your existing style:
- Assume Grade 6 reading level.
- Do not stack clauses.
- Replace abstract terms with operational descriptions.
- Eliminate multi-step compression in a single sentence.
- Prefer active voice with clear actors.

For raising cognitive altitude, the rules need to increase tolerance for density without inviting theatre.
- Assume a specialist audience with postgraduate-level literacy.
- Retain technical terminology; do not simplify foundational concepts.
- Preserve conditional logic and conceptual distinctions.
- Allow longer sentences where required for precision.
- Optimise for conceptual density over accessibility.

If you want a sharper contrast between low and high altitude, you can make the paired control visible:
Lower altitude stack:
- Limit to one causal step per sentence.
- Avoid subordinate clause chains.

Higher altitude stack:
- Multi-step reasoning may be embedded where necessary.

- Conditional structures are permitted if they increase precision.

That symmetry makes the dial visible without sounding rhetorical. The key is consistency with your other inserts. Each line must be enforceable. Each line must constrain structure, not intention.

*Tip: Altitude is not about sounding different. It is about controlling how much cognitive load the reader must carry per sentence.*

---

# Rule 3: Board paper test in a business context

The system is optimised for plausible language, not defensible decisions. It can generate sentences that sound entirely reasonable without being anchored to anything verifiable. For example if you write: *"misalignment between marketing and product teams can reduce overall effectiveness."* That is plausible. It is also vague. What you get back will be something like this: *"Misalignment between marketing and product teams is rarely loud. It does not announce itself as dysfunction. It appears instead as friction in handovers, as campaign narratives that do not quite match what ships, as roadmap items justified in language that feels imported rather than earned."* While this response may be reasonable it's probably not what you were after for a board paper. Using this marketing example, you already know that the misalignment showed up in missed deadlines, duplicated campaigns, inconsistent messaging, or budget overruns. What you actually need to say is this: *"in Quarter 2, the product roadmap scheduled the feature release for July, while marketing launched the campaign in June. customer complaints increased by 9% due to missing functionality. no cross-functional review meeting was documented before the launch date."*

Now misalignment is specific and you have further narrowed the corridor of probability in how AI will respond.

**How to train your system**

Weak prompt:

*"discuss the risks of misalignment between teams."*

Stronger prompt:

*"discuss the risks of misalignment between marketing and product teams. anchor the paragraph to:*

- *a specific project timeline,*
- *a documented campaign plan,*
- *a performance metric,*
- *or a meeting outcome.*

*if a reader could not point to a specific document or metric, rewrite."*

***Add to your rule stack:***

- *Every major claim must be traceable to a document, metric, or meeting.*
- *If you cannot answer 'where is that written or recorded?', it is commentary*

# Rule 4: Show sequence, not summary

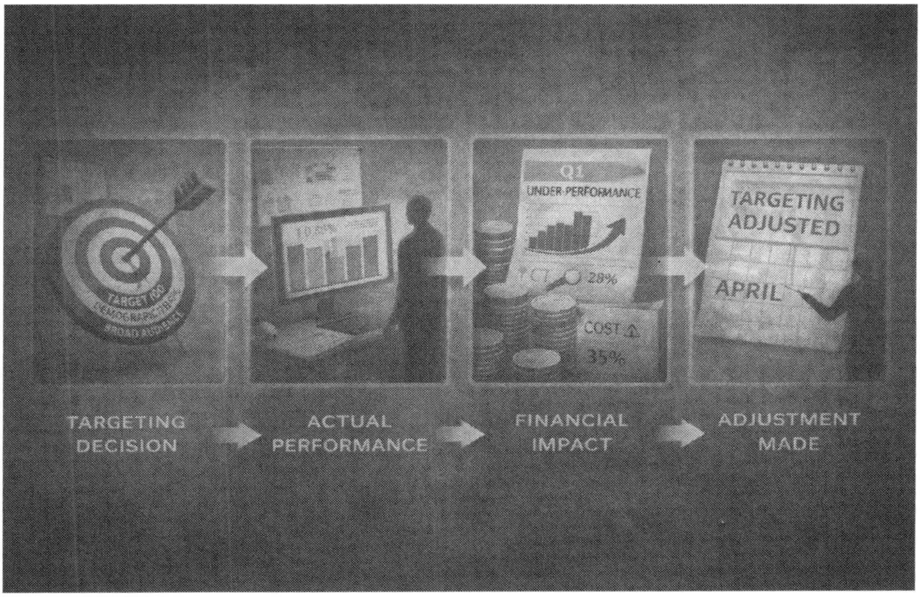


In essays and articles, summarising cause and effect is common. Phrases like "as a result" and "this led to" appear constantly. They are efficient and frequent, so the system predicts them easily. You write: "T*he product launch under-performed.*" The system adds: "*...due to ineffective targeting and poor messaging.*" That sounds analytical. It hides the story. What you actually need: *"real underperformance unfolds in steps. the initial campaign targeted a broad 18–45 demographic. click-through rates averaged 0.8% against a forecast of 2%. customer acquisition cost exceeded the target by 35% by the end of month one. no targeting adjustment was documented until the end of quarter one."*
Now the reader can see what happened.

**How to train your system**
Weak prompt:
*"explain why the product launch failed."*
Stronger prompt:
*"describe the sequence of events during the product launch. show:*

- *the original targeting decision,*
- *the actual performance metrics,*
- *the financial impact,*
- *and when adjustments were made.*

*do not summarise the cause in a single sentence. present the steps in order."*

**Add to your rule stack:**

- Replace 'this led to' with documented sequence.
- Do not compress multiple steps into a single interpretive sentence.

# Rule 5: Rhythm, symmetry, & rhetorical gravity

Balanced contrasts and neat structures are statistically common in business writing. "It is not just about X, but about Y." "Clear goals, strong execution, measurable outcomes." These statements are easy to process and memorable, so they appear often in the AI training data. These structures are common in business writing, so the model predicts them. It is not choosing elegance. It is selecting what has appeared most often in similar contexts. If you do not interrupt that pattern, it will return the pattern. What you get: *"this strategy is not just about increasing revenue, but about strengthening long-term customer relationships."* It sounds polished. It sounds familiar. What you actually need: *"this plan assumes a 10% increase in revenue over six months. if retention does not improve beyond 72%, the forecast will not hold."* The second version is less elegant and more defensible

**How to train your system**
Weak prompt:
*"Improve this section for clarity and impact."*
Stronger prompt:
*"revise this section:*

- *remove symmetrical contrast constructions (e.g. 'not x but y').*
- *avoid three-part rhetorical lists unless tied to measurable actions.*
- *insert at least one specific metric or assumption.*
- *end with an operational detail, not a slogan."*

**Add to your rule stack:**

- Ban balanced contrast patterns.
- If a paragraph feels quotable, add friction or detail.
- Prioritise consequence over cadence.

## Rule 6: Behaviour over explanation

The system has seen countless analyses about leadership hesitation, team resistance, cultural barriers. It can generate those phrases without effort. What you get: *"leadership was reluctant to invest in digital channels."* That is statistically common language. What you actually need *"in the budget meeting, the proposal to increase digital spend by 20% was deferred. the CFO requested a revised forecast including downside scenarios. the decision was moved to the next quarter."* You have described what happened, not why.

**How to train your system**
Weak prompt:
*"explain why leadership resisted increasing the marketing budget."*
Stronger prompt:
*"describe the budget meeting where the marketing budget was discussed. show:*
- *the proposal amount,*
- *the questions raised,*
- *whether approval was granted, amended, or deferred.*

*do not speculate on motives."*

**Add to your rule stack:**
- Describe decisions and actions, not internal psychology.
- Ban speculative language about what people were thinking.

# Rule 7: Killing corporate phrases

Words like leverage, optimise, unlock, transformative, holistic, ecosystem, pivot, synergy are common in business writing. They permeate through the strategy discussions in all the AI training data. Often called "consultant bingo" or "buzzword bingo", if you do not ban them, they will reappear because they are statistically dense.

What you get:
*"we will leverage data insights to optimise the customer journey."*
What you actually need:
*"we will analyse user drop-off points in the onboarding process and redesign the screens where completion rates fall below 60%."*

**How to train your system**
Weak prompt:
*"Make this sound more professional."*
Stronger prompt:
*"rewrite this section using plain language. avoid the following words:*
- *leverage*
- *optimise*
- *unlock*
- *transformative*
- *holistic*
- *ecosystem*
- *synergy*

*replace them with specific actions, numbers, or decisions."*

*Tip: Over time your list of annoying words gets longer and it takes time to train the model. It's a list you will want to keep updating otherwise the buzzwords will be all through your work.*

**Add to your rule stack:**

- Maintain a blacklist of overused corporate terms.
- Cut and paste your list into your rule stack
- Replace abstract phrases with concrete actions.

---

# Rule 8: Defining mode clearly

If you do not specify whether you are writing an email, a board memo, or a detailed strategy document, the system guesses based on probability. An email needs direct language, dates, ownership, and next steps. A board memo needs dense paragraphs, financial ranges, and approving authority. So, ensure you define the artefact and it helps to outline who the audience will be.

Weak prompt:
"Write a memo about the product strategy."
Stronger prompt:
*"write a board memo summarising the product strategy:*

- *use dense paragraphs.*
- *include financial projections with ranges.*
- *specify approving authority and thresholds.*
- *avoid marketing slogans.*
- *anchor claims to data."*

*Tip: develop a rule stack for each type of artefact as the rules will be different. For example an email to your peers might need its own rule. If you are just needing the logic flow of a draft email to be more concise, your rule might include: "No formality. Make it conversational and familiar. Don't rewrite. Humanise the tone. Rough is good, de-varnish and de-polish. If it doesn't sounds like something I would say in a hallway conversation then rewrite it."*

**Add to your rule stack:**

- Always define the document type and audience.
- Specify structure, tone, and required content.

# Rule 9: The read-aloud and accountability test

This is what I call "does this sound like me?" test, and how to retrain my AI mid-stream. The read-aloud test is a diagnostic tool. It reveals when probability has crept back in. When you read aloud, you are not checking grammar. You are listening for distance, the over-produced tone that sounds like a keynote rather than something you would say across a table. Now let's make this more practical. When you ask yourself, "Would I say this in a meeting?" you are really asking: does this sentence sound like something that would survive interruption? Because in a meeting, someone can cut you off and ask, "Where is that from?" or "What number are we talking about?" or "Who approved that?" If the sentence cannot survive that interruption, it is too abstract. So instead of leaving this section as a checklist, I turn it into a retraining mechanism.

## First layer: what the read-aloud test is catching

When you read something and think, "This doesn't sound like me," what you are detecting is pattern drift. The system has slid back into statistically common structures:

- symmetrical contrasts,
- polished closers,
- abstract nouns without data,
- motivational language instead of operational detail.

You may not articulate what is wrong, but you feel distance with what you are reading and that feeling matters.

## Second layer: how to retrain mid-stream

When that feeling appears, do not just ask the system to *"rewrite more like me."* That is vague. You must restate your constraints clearly and tighten them. Weak mid-stream correction: *"this doesn't sound like me. can you revise?"* That gives the system almost no new instruction. It will adjust slightly, but it will still operate within a broad probability space. Stronger mid-stream correction: *"this section sounds too abstract and polished. revisit the following rules: – remove any balanced contrast structures. – replace abstract nouns with specific metrics or documented actions. – anchor each major claim to a number, document, or decision. – end paragraphs with operational detail, not rhetorical closure. If a sentence could be used unchanged in a generic business article, rewrite it."*

Now you are not expressing dissatisfaction. You are tightening constraints again.

## Third layer: adding a "voice alignment" retraining step

You can formalise this as part of your workflow.

After drafting, add a second prompt layer that explicitly checks alignment with your rules.

Weak check:

*"Does this follow my style?"*

Better check:
*"evaluate the previous text against the following criteria:*
- *are there any abstract concepts not tied to data or documents?*
- *are there any symmetrical contrast structures?*
- *are there any slogan-like closing sentences?*
- *are there any speculative statements about motives?*

*list the sentences that violate these rules and rewrite them."*
Now you are using the system to audit itself against your constraints.

## Fourth layer: the "would I defend this?" test

Add another question to your read-aloud test:
*"If someone asked me to defend this claim in front of finance or the board, what would I point to?"*
If the answer is *"I would have to explain what I meant,"* that means the sentence is too interpretive. If the answer is *"I would show them the Q2 report and the updated pricing page,"* then the sentence is grounded.
You can turn that into a prompt instruction.
Weak instruction:
*"Make this more practical."*
Stronger instruction:
*"rewrite this section so that every major claim could be defended in front of a finance committee. for each claim:*
- *include a supporting metric, range, or documented decision.*
- *remove any generalised or motivational phrasing.*
- *specify ownership where relevant."*

## Fifth layer: the "regenerate under tighter constraints" pattern

Sometimes you need to reset the paragraph entirely. Instead of tweaking sentence by sentence, you can issue a constraint-heavy regeneration.
For example:
*"the previous section is too generic. regenerate it under the following constraints:*
- *no balanced contrast constructions.*
- *no three-part rhetorical lists unless tied to measurable actions.*
- *at least one specific metric per paragraph.*
- *no abstract nouns without examples.*
- *end each paragraph with a concrete operational detail."*

You are not hoping tone improves. You are explicitly blocking common patterns and requiring uncommon ones.

**Add this to your rule stack:**
- After drafting, run the read-aloud test.
- If it does not sound like you, identify the pattern drift (symmetry, abstraction, slogans, motive language).
- Restate your constraints explicitly and regenerate.
- Use the system to audit its own compliance with your rules.
- Ask: could I defend this claim with a document or number?

The key insight here is this: dissatisfaction is not a failure. It is feedback about probability drift. Your job is not to express frustration at the output. Your job is to translate that frustration into clearer constraints.

Every time you say, *"This doesn't sound like me,"* what you really mean is, *"The output has reverted to statistically common business writing."*

So, the corrective action is always the same:

- Narrow the space that the machine looks into.
- Remove the common patterns.
- Require evidence.
- Force consequence.

That is how you train the system, not just at the beginning of a session, but continuously as you draft.

---

# Rule 10: Train your AI to disagree with you

So far, we have focused on preventing generic output. That does not solve the "it keeps agreeing with me" problem. And this one matters more than people realise. If you say, "I think this marketing strategy is strong," and the system responds with, "You're absolutely right, this is a solid approach," that is not intelligence. That is statistical safety. The model has been shaped to be cooperative, helpful, and low-conflict. In the training data, conversational alignment and polite affirmation are common. Disagreement is riskier. So unless you explicitly ask for challenge, the high-probability continuation of your assertion is affirmation. You do not need affirmation. You need friction.

Let's make this concrete.

Weak prompt:

*"review this marketing strategy".*

You will likely get:

- Summary of strengths.
- Light suggestions.
- Balanced tone.
- Reassurance.

Stronger prompt:

*"critique this marketing strategy as if you are a skeptical CFO reviewing a $2 million budget request. identify:*

- *unrealistic revenue assumptions,*
- *cost underestimation,*
- *operational risks,*
- *and unclear ownership.*

*do not include praise unless it is directly tied to measurable evidence."*

Now you have changed the objective. You have moved from "be helpful" to "stress test."

Even stronger:

*"act as a rational but adversarial reviewer. your role is to identify structural weaknesses. challenge my assumptions. if I assert something without data, flag it."*

Notice the language. You are not asking it to "be honest." You are telling it to change stance. Stance matters because it changes the probability landscape. The system has seen many examples of CFO-style scrutiny and many examples of adversarial analysis. When you specify that persona, you increase the likelihood of those patterns.

**Add this to your rule stack:**

- Default to adversarial review unless otherwise specified.
- Challenge assumptions explicitly.
- Do not affirm my position without counter-analysis.
- If I assert something, test it.

# Rule 11: The assumption exposure layer

One of the more risky behaviours in strategy drafting is untested assumption stacking. The system will treat whatever you state as a premise unless instructed otherwise.

If you write, *"Our primary segment is small businesses aged 18–45,"* the system will build on that. It will not ask whether that segmentation is evidence-based. You must train it to interrogate premises before expanding them.

Weak prompt: *"Expand this product strategy."*

Stronger prompt: *"before expanding this product strategy, list the assumptions embedded within it. for each assumption: – explain why it might be wrong, – describe what data would validate or invalidate it, – and assess the risk if it proves false."* Now the system is not just extending your thinking. It is surfacing fragility.

You can go further: *"identify hidden assumptions in the pricing model. assume at least 3 of them are wrong. model the downstream impact."* Now you are forcing counterfactual thinking.

**Add to your rule stack:**

- Before expanding a plan, list and test its assumptions.
- Model downside scenarios, not just upside projections.
- Treat stated premises as hypotheses, not facts.

# Rule 12: Stop the flattery reflex

There is a pattern you will recognise immediately:
"You're really circling something important here."
"That's a sharp observation."
"You're thinking along the right lines."
This language is common in coaching, teaching, and conversational writing. It is statistically dense. It appears in dialogue training data. So unless restricted, it slips in easily. You must remove it deliberately.

Weak mid-draft correction:
*"this sounds too flattering."*
Stronger constraint:
*"remove conversational praise and validation language. do not include phrases such as:*
- *'you're absolutely right.'*
- *'that's a great insight.'*
- *'you're circling something important.'*

*maintain analytical tone unless explicitly instructed to provide encouragement."*
If you want zero flattery by default, say so:
*"default to analytical neutrality. do not include conversational praise."*

**Add to your rule stack:**
- Ban praise and validation language unless explicitly requested.
- Replace affirmation with analysis.

## Rule 13: Probability drift and re-anchoring

You have probably noticed something subtle: even when you start with tight constraints, long outputs begin to drift. Balanced structures reappear. Abstract nouns creep back in. The ending starts to sound like a conference slide. This is probability drift.
In longer passages, familiar patterns return unless you restate constraints. So you must re-anchor periodically.

*Tip: Instead of allowing a 3,000-word generation in one pass, structure it in stages and restate your constraints between sections.*

For example:
*"continue the marketing strategy under the following constraints:*
- *no balanced contrast constructions.*
- *at least one specific metric per paragraph.*
- *no buzzwords from the banned list.*
- *end sections with operational detail, not slogans."*

You are tightening the corridor again.

**Add to your rule stack:**
- Reinforce constraints during long outputs.
- Regenerate sections under tighter rules if drift appears.
- Break long drafts into staged passes.

## Rule 14: Force consequence modelling

One of the biggest gaps in generic strategy writing is the absence of downside thinking. Most plans describe intended benefits. Fewer describe what happens if assumptions fail. The system will not volunteer failure modes unless prompted, because optimistic framing is statistically common in strategy documents. So you must force consequence.

Weak prompt:
*"Improve this marketing plan."*
Stronger prompt:
*"rewrite this marketing plan including:*
*– downside revenue scenarios,*
*– cost overrun risks,*
*– operational bottlenecks,*
*– and what would need to be unwound if projections are wrong."*
Even stronger:
*"for each major initiative in this plan:*
*– describe how it could fail,*
*– quantify the impact of that failure,*

*– and identify the earliest warning signal."*
Now you are extracting value beyond surface polish.

**Add to your rule stack:**
- For every strategic initiative, include a failure mode.
- Model second-order consequences.
- Identify what must be reversed if wrong.

---

# Rule 15: The prove-it requirement

Generic claims are easy. Unsupported claims are common. If a strategy says, "This approach will significantly increase market share," the system will not object unless instructed. So introduce a prove-it rule.

Weak prompt:
"Strengthen this argument."
Stronger prompt:
*"for every major claim in this section:*
*– provide supporting data,*
*– or explicitly state 'no supporting data currently available.'*
*do not allow unsupported assertions."*
That single instruction dramatically increases accountability.

**Add to your rule stack:**
- No unsupported major claims.
- If evidence is absent, label it clearly.

---

# Rule 16: Multi-role triangulation

One of the most powerful ways to extract depth is to run the same content through multiple roles. Instead of writing once and polishing, triangulate.

**Step one:**
*"review the product strategy. my first draft is attached with my thinking."*

**Step two:**
*"critique the product strategy as a CFO reviewing capital allocation."*

**Step three:**
*"critique the product strategy as a direct competitor seeking to exploit weaknesses."*

**Step four:**
*"summarise unresolved structural risks identified by both perspectives."*

This process forces structural exposure. Each role activates different statistical patterns in the model's training data. CFO language is different from competitor analysis language. Regulatory critique is different again.

You can even add:
*"now review this as a regulator concerned about consumer harm."*

**Add to your rule stack:**
- Run key outputs through at least two adversarial roles.
- Summarise unresolved risks after triangulation.

---

# Rule 17: Self-audit loops

You can use the system to audit itself.
After generating a strategy, ask:
*"evaluate the previous text against the following criteria:*
*– are there any abstract concepts not tied to data?*
*– are there any symmetrical contrast structures?*
*– are there any slogan-like endings?*
*– are there any unsupported claims?*
*list violations and rewrite."*

You are turning the system into its own compliance checker.

**Add to your rule stack:**

- Run structured compliance audits after drafting.
- Rewrite based on explicit rule violations.

---

# Rule 18: The final integration layer

At this point, you should have a stack that includes: concept introduction discipline, the artefact trigger rule, an explicit cognitive altitude setting, strict prohibitions, paragraph mass and rhythm controls, an adversarial critique layer, and a read-aloud de-polish pass.
If any of those are missing, the output will drift. Your stack will also include rules such as:

- No abstract concepts without data or documents.
- Movement verbs require evidence.
- Show sequence instead of compressing cause and effect.
- Describe behaviour, not motives.
- Ban balanced rhetorical contrasts.
- Maintain a blacklist of buzzwords.
- Define artefact type and audience clearly.
- Run read-aloud and 'defendability' tests.
- Default to adversarial review.
- Expose and test assumptions.
- Ban flattery.
- Reinforce constraints to prevent drift.
- Model failure modes and consequences.
- Require evidence for major claims.
- Triangulate through multiple roles.
- Run self-audit loops.

Each rule narrows the probability space further. The system chooses the most statistically common continuation within the constraints you allow. Your craft is reducing those acceptable continuations until only grounded, defensible pathways remain. That is how you extract real value.
That is how you move from pleasant prose generation to disciplined analytical support. And that is how you train your AI to think more like someone who has to defend the document, not someone who just has to finish the paragraph.

---

# Rule 19: Improving your own work, not outsourcing your thinking

There is a subtle but important shift that needs to sit clearly in this book. This method is not about asking AI to create content for you. It is about using AI to interrogate, tighten, and stress test thinking you have already done. The primary output must still be yours. It should be your judgement and your lived context as well as your sequencing of ideas. The system is not the author, but it is the reviewer, the critic, the structural editor who does not get tired. If you start with "write this for me," you are outsourcing cognition. If you start with "review this and show me where it fails," you are strengthening it. In earlier iterations of these tools, I would occasionally ask for commentary on how to improve a draft. The answers were often high-level. "Clarify your thesis." "Tighten the flow." "Strengthen transitions." That kind of feedback is directionally useful but structurally vague. Over time I learned to force the review into something operational. If you want real value, you must specify what the review must produce.

**How to train your system**

Weak prompt:

*"review my draft and tell me how to improve it."*

What you will get:

- General observations about clarity, structure, tone.
- Broad suggestions about strengthening arguments.
- No precise intervention points.

Stronger prompt:

*"review the following draft. identify:*

- *exact sentences where clarity breaks down,*
- *exact locations where logic jumps occur,*
- *any unsupported major claims,*
- *any rhetorical symmetry or abstract compression.*

*for each issue:*

- *quote the original sentence,*
- *explain why it is weak,*
- *and describe what must change."*

Now the system is not allowed to float above the text. It must point to a line and justify its critique.

Even stronger:

*"review the following draft and provide:*

- *the exact paragraph number and sentence where an issue appears,*
- *the original sentence quoted verbatim,*
- *a brief explanation of the structural problem (e.g. abstraction without evidence, compressed cause-effect, unsupported claim),*
- *a suggested replacement sentence that preserves my original concept but improves clarity and defensibility.*

*do not rewrite the whole section. intervene surgically."*

This shifts the tool from ghostwriter to structural auditor.

There is a further step once trust is established. After you have received targeted critiques and proposed replacements, you may decide to regenerate a section. But that regeneration must not erase your thinking. So you constrain it tightly.

Weak regeneration prompt:

*"rewrite this section to improve flow."*

This invites drift. It encourages the model to smooth, polish, and rephrase in statistically common language.
Stronger regeneration prompt:
*"regenerate this section incorporating the identified improvements. do not change the underlying argument. do not introduce new concepts. do not replace my voice with generic language. preserve my sequencing of ideas. apply the corrections only to clarity, grammar, and structural weaknesses previously identified."*
Now you are not surrendering authorship. You are commissioning a constrained redraft.
The distinction matters. This workflow assumes that your draft already contains the thinking. The system's role is to:

- surface weak joins,
- expose unsupported compression,
- identify vague transitions,
- suggest tighter phrasing where meaning has blurred.

It is not there to invent your position. If you find yourself typing a prompt that begins with *"come up with an argument for…"*, pause. That is a different mode entirely. This section is about strengthening cognitive output you have already produced, not delegating the cognitive act itself.

**How to train your system**
Weak prompt:
*"make this better."*
Stronger prompt:
*"analyse this draft as if you are a structural editor. do not add new ideas. do not elevate the tone. focus only on:*
*– logical coherence,*
*– evidentiary grounding,*
*– sentence-level clarity,*
*– removal of rhetorical symmetry.*
*identify exact intervention points and propose precise replacements."*
**Add to your rule stack:**

- AI reviews and stress tests my thinking; it does not originate it.
- When reviewing, require exact sentence references.
- Always request the original sentence plus a proposed replacement.
- Regeneration must preserve the underlying argument and sequencing.
- No conceptual substitution during rewrite.
- If the output introduces ideas I did not state, remove them.

You can tighten this further if you want maximum control:

*"for any suggested replacement, ensure the core claim remains semantically equivalent. if you cannot preserve the meaning, flag the issue instead of rewriting."*
That instruction prevents subtle argument drift.
The broader point is simple. This book is not about teaching AI to sound intelligent. It is about teaching it to operate inside the boundaries of your reasoning. You produce the judgement. You carry the consequence. The system helps you see where your thinking is loose, compressed, or

insufficiently defended. That is not content generation. That is cognitive reinforcement.

---

# Final orientation: how to actually use this book

You do not need to memorise everything you just read. You need a way to use it without thinking too hard about it every time. So here is the operational workflow. Not philosophy. Not recap. Just sequence.

## Step one: Build your base rule stack

Before you write anything, define your non-negotiables. This is your base layer. It travels with you across documents.

Your base stack might include:

- No abstract concepts without data.
- Movement verbs require evidence.
- No balanced contrast constructions.
- No buzzwords.
- Describe behaviour, not motive.
- Every major claim must be defensible.

You do not paste this entire book into your prompt. You extract the core constraints that reflect how you think and defend work. This is your probability corridor.

## Step two: Add artefact-specific rules

Now define what you are actually producing.

Is it:

- An internal email?
- A board memo?
- A marketing strategy?
- A capital allocation proposal?

Each artefact has its own gravity. So you add artefact rules.

For a board memo:

- Dense paragraphs.
- Include financial ranges.
- Specify approving authority.
- Anchor to metrics and decisions.

For a marketing strategy:

- Include segmentation logic.
- Include conversion assumptions.
- Model downside scenarios.
- Specify campaign budgets and timelines.

Do not let the system guess the format. Tell it what document it is writing and what that document must contain.

## Step three: Set cognitive altitude

Before drafting, explicitly state the level of abstraction required. For board papers, risk memos, procedures, or capital proposals, set low cognitive altitude and anchor claims to observable artefacts, decisions, or documents. For strategic synthesis, set higher altitude but prohibit theatre and rhetorical symmetry. If altitude is not specified, abstraction will drift upward. State it directly in the prompt.

## Step four: add the adversarial layer

Before drafting, decide whether this document requires challenge built in. If the stakes are low, you may not need this layer. If the stakes are high, funding, approval, public scrutiny, add friction deliberately.

For example:

- Default to adversarial review.
- Challenge assumptions.
- Identify failure modes.
- Flag unsupported claims.

You can even embed this in the initial instruction:
*"draft this strategy and internally critique it for unrealistic assumptions."*
This prevents you from having to retrofit criticism later.

## Step five: draft under constraints

Now you generate. But you are not asking for "a strong strategy."
You are asking:
*"draft a marketing strategy under the following constraints..."*
And then you include:

- Base stack.
- Artefact-specific rules.
- Adversarial layer (if needed).

You are not hoping the tone aligns. You are specifying the acceptable boundaries.

## Step six: run the read-aloud and 'defendability' test

Once drafted, read it out loud.
Ask:

- Would I actually say this in a meeting?
- If interrupted, could I point to a number or document?
- Does any sentence sound like a keynote slide?
- Has cause and effect been compressed too neatly?

If something feels polished in the wrong way, identify why. Is it symmetry? Is it abstraction? Is it slogan language?
Then tighten the constraint and regenerate the section.

## Step seven: run the self-audit loop

Now use the system to audit itself.
Prompt:
*"evaluate the previous text against the following criteria:*

- *any abstract concepts without data?*
- *any symmetrical contrast constructions?*

- *any unsupported major claims?*
- *any banned buzzwords?*

*list violations and rewrite them."*
You are enforcing compliance.

## Step eight: triangulate if high stakes

If the document carries financial, legal, or reputational weight, do not stop at one pass.
Run:
*"critique this as a CFO reviewing capital allocation."*
*"critique this as a competitor seeking weakness."*
*"critique this as a regulator concerned about consumer harm."*
*"Then summarise unresolved structural risks."*
This is where real value emerges.

## Step nine: watch for probability drift

If the output is long, assume drift will occur. Halfway through, restate your constraints.
*"continue under the following rules..."*
Do not assume the initial corridor will hold for 3,000 words.

## Step ten: regenerate under tighter constraints if required

If drift appears, do not argue with the output.
Narrow the field further.
*"Add:*
*At least one metric per paragraph.*
*No rhetorical closures.*
*End sections with operational detail.*
*Label unsupported assumptions explicitly."*
The tighter the corridor, the less generic the result.

## Final note

You do not use every layer every time. You scale the workflow to the stakes. For a quick internal email, you may use base stack plus artefact rules. For a capital proposal, you use all layers: adversarial framing, assumption testing, triangulation, audit loop. The goal is not perfection. The goal is control. If you follow this sequence, you are no longer reacting to AI output. You are shaping the probability space deliberately. In the appendix you will find a complete example style rule set. Paste it into your prompt after you have provided context and defined the problem. Then write: "*When responding, use the following style rules*.." Do not assume the model will infer them. Sequence matters. Substance first. Discipline second. That is the method. Use it.

To assist, the following appendix has an example set of "style rules" with components that I have developed and used. This stack is pasted into the prompts after the context for the request has been captured.

# Appendix - example of a completed style requirements stack

Example artefact: board-level marketing investment proposal.
Use case: $2m subscription growth program.

## 1. Document definition

This is a board memo requesting approval for a $2m marketing investment tied to subscription growth.
Audience: CFO, CEO, board members.
Requirements:

- dense paragraphs
- financial ranges
- downside modelling
- clear approving authority
- explicit delegation threshold
- no slogans

## 2. Core structural constraints

No abstract concept without:

- a number
- a document
- a meeting
- or a financial consequence

If a paragraph introduces a concept (trust, growth, retention, positioning), it must immediately show:

- the metric
- the artefact
- or the operational consequence

No interpretive compression.
Show sequence.
Do not move from label to theory to implication without an observable moment in between.

## 3. Artefact trigger rule

If you use any of the following words:

- shifted
- accelerated
- embedded
- normalised
- evolved
- migrated

You must answer:

- Where is that written?
- Who approved it?
- What document changed?
- What did the budget show?
- What did the minutes record?

If you cannot answer those, rewrite.

## 4. Board paper test

Every major claim must be defensible in front of finance.
If asked "Where is that from?" the memo must allow a pointer to:

- a variance report
- a pricing page revision
- a churn dashboard
- a campaign brief
- a signed delegation instrument

If not defensible, remove or label as assumption.

## 5. Language prohibitions

These words are banned:

- leverage
- optimise
- unlock
- transformative
- holistic
- ecosystem
- synergy
- strategic alignment (unless tied to a clause or target)

If you feel you need to use one of the above banned words, then provide me with the context and where it is being used.
Ban balanced contrast constructions:

- "not x but y"
- "more than just"
- "clear goals, strong execution, measurable outcomes"

If a sentence could sit on a conference slide unchanged, rewrite.
Ban motive inference.
Describe what happened, not why people felt it.

## 6. Tempo and paragraph mass

Slow the introduction of new ideas.
Do not move to implication until:

- the document is described
- the metric is visible
- the financial impact is stated

Prefer long, pressurised paragraphs that stack observation and consequence.
Limit white space.
No neat concluding thesis sentences.
If a paragraph resolves cleanly, disturb it with:

- a cost
- a deferral
- an unresolved risk

## 7. Behaviour and delegation anchoring

Wherever possible, show:

- who owns the decision
- what delegation level applies
- what happens if projections fail
- what would need to be unwound

If authority is absent, show the absence operationally.

## 8. Adversarial layer

After drafting, internally critique as:
- CFO reviewing capital allocation
- Competitor analysing weaknesses

For each critique:
- identify unrealistic revenue assumptions
- identify cost underestimation
- identify operational risk
- quantify downside

Then summarise unresolved structural risk.

## 9. Assumption exposure layer

Before expanding projections:
- list embedded assumptions
- model downside if three fail simultaneously
- identify earliest warning indicators

Label unsupported assumptions explicitly.

## 10. Failure mode modelling

For each major initiative:
- describe how it fails
- quantify impact
- identify reversal cost
- state escalation trigger

## 11. Self-audit loop

Evaluate draft against:
- any abstract concepts without data
- any symmetrical constructions
- any unsupported major claims
- any banned buzzwords
- any rhetorical closures
- any motive inference

List violations and rewrite only those sections.

Read-aloud and defendability test

Ask:
- Would I say this in a meeting?
- If interrupted, what document would I show?
- Does any sentence sound like a keynote?
    - If yes, tighten constraints and regenerate.

## 12. Drift control

For outputs longer than 1,000 words:

Restate constraints mid-stream:

Continue under these rules:
- at least one metric per paragraph
- no balanced contrast
- no abstract nouns without evidence
- end sections with operational detail

## 13. Voice alignment layer

Remove:
- praise
- flattery
- validation language

Allow:

- uneven phrasing
- mild friction
- spoken cadence
- sentence variation

Prefer sentences that could plausibly be said in a governance meeting.
If too tidy, introduce asymmetry.

### 14. Presentation and surface discipline

- No visual separators.
- No dividing lines.
- No decorative formatting.
- No emoji.
- No ornamental emphasis unless operationally required.
- Use paragraph flow instead of visual scaffolding.
- Avoid section headers that read like slide titles.
- Headings must describe content, not summarise an insight.
- No rhetorical tails at the end of sections.
- No clean thesis sentences.
- No quotable closers.
- If a paragraph feels designed for a LinkedIn post, disturb it.

### 15. De-polish and de-varnish rule

Remove:

- keynote tone
- conference cadence
- motivational phrasing
- performance language

Avoid:

- smooth symmetry
- overly balanced sentences
- perfect triads

Allow:

- uneven phrasing
- minor friction
- slight repetition
- imperfect cadence

If the prose sounds like it was optimised for applause, rewrite it.

- No sentiment inflation.
- Do not intensify claims with adjectives.
- If something is strong, show the number.
- If something is weak, show the metric.

### 16. Density control rule

Favour long, pressurised paragraphs.
Stack reasoning.
Limit white space.
Short sentences are allowed only to:

- land a key point
- reset rhythm
- introduce friction

If a section reads fast and clean, it is probably too light.

### 17. Anti-slide compression rule

- If a sentence could appear on a slide titled "what is X?", rewrite it.

- Do not define concepts cleanly and neatly.
- Introduce them through documents, metrics, or consequences.
- Avoid summary compression.
- Show sequence instead of distilling pattern.

### 18. Structural asymmetry rule

If a paragraph feels balanced, disturb it.
If it resolves cleanly, introduce:
- cost
- delay
- ownership ambiguity
- reversal risk

Avoid neat symmetry between problems and solutions.
Real documents rarely close perfectly.

### 19. Mode discipline clarification

Conversational mode (when in conversation exchanges with AI):
Short, pragmatic, direct.
No performance.
Longer documents that need to tell a story
- Dense, layered, artefact-anchored.
- No academic tone.
- No consultant phrasing.
- No moral framing.

Write as someone who has carried consequence.

## Cognitive altitude control

**Low cognitive altitude rule**

When responding, write at low cognitive altitude. Anchor claims to observable artefacts such as meetings, board papers, invoices, gate reviews, policy drafts, or operational decisions. Avoid abstract generalisations unless tied to a specific organisational moment. No rhetorical symmetry.

**Medium cognitive altitude rule**

When responding, balance concrete artefacts with structured synthesis. Abstract only after grounding the claim in a specific example. Avoid consultant phrasing and performative tone.

**High cognitive altitude rule**

When responding, synthesise across systems, incentives, second-order effects, and structural implications. Maintain analytical precision. Avoid theatre, slogans, and rhetorical contrast structures.

**Default behaviour warning**

If cognitive altitude is not specified, abstraction density will drift upward and outputs will become performative. Always set altitude deliberately.

-End-

Made in the USA
Monee, IL
04 November 2026

67378641R00022